ULTIMATE SPECIAL FORCES

US AIRBORNE FORCES

TIM COOKE

FARGO PUBLIC LIBRARY

PowerKiDS press

New York

Published in 2013 by The Rosen Publishing Group, Inc.
29 East 21st Street, New York, NY 10010

Copyright © 2013 by Brown Bear Books Ltd

All rights reserved. No part of this book may be reproduced in any form without permission from the publisher, except by a reviewer.

Senior Editor: Tim Cooke
US Editor: Sara Antill
Designer: Supriya Sahai
Creative Director: Jeni Child
Picture Researcher: Andrew Webb
Picture Manager: Sophie Mortimer
Children's Publisher: Anne O'Daly
Editorial Director: Lindsey Lowe

PHOTO CREDITS
Front Cover: US Department of Defense
Library of Congress: 17br, 21, 22/23; Robert Hunt Library: 25, 45c; US Air Force: 33, 38, 40, 41t; US Army: 4, 5, 7, 08, 9, 11, 13t, 13br, 15, 17bl, 18, 19, 34, 35, 42bl, 43tr; US Department of Defense: 6, 10, 12, 14, 16, 20, 24, 26, 27, 28, 29, 30, 31, 32, 36, 37, 39, 41bl, 44, 45tr.
Key: t = top, c =center, b = bottom, l = left, r = right.

Library of Congress Cataloging-in-Publication Data

Cooke, Tim, 1961–
 US Airborne forces / by Tim Cooke.
    p. cm. — (Ultimate special forces)
Includes index.
  ISBN 978-1-4488-7882-6 (library binding) — ISBN 978-1-4488-7959-5 (pbk.) —
ISBN 978-1-4488-7964-9 (6-pack)
1. United States. Army—Airborne troops—Juvenile literature. 2. United States. Army—Parachute troops—Juvenile literature. I. Title.
  UD483.C66 2013
  356'.1660973—dc23

2012012447

Manufactured in the United States of America

CPSIA Compliance Information: Batch #B2S12PK: For further information, contact Rosen Publishing, New York, New York, at 1-800-237-9932

CONTENTS

INTRODUCTION 4
ORGANIZATION 6
82nd AIRBORNE 8
101st AIRBORNE 10
AIR ASSAULT SCHOOL 12
JUMP SCHOOL 14
PARACHUTE DROPS 16
AIR ASSAULT 18
WORLD WAR II 20
BASTOGNE 22
VIETNAM 24
GRENADA 26
PERSIAN GULF 28
IRAQ 30
AFGHANISTAN 32
HURRICANE KATRINA 34
HAITI 36
AIRCRAFT 38
HELICOPTERS 40
SPECIAL GEAR 42
WEAPONS 44
GLOSSARY 46
FURTHER READING AND WEBSITES 47
INDEX 48

US AIRBORNE FORCES

INTRODUCTION

Since airplanes became part of warfare in World War I (1914–1918), commanders have wanted to drop soldiers from the sky. Aircraft can quickly get troops deep behind enemy lines. In World War II (1939–1945) thousands of US paratroopers were dropped into enemy territory.

MEMBERS OF THE 101st Airborne cross the English Channel before jumping over Normandy on D-Day, June 6, 1944.

INTRODUCTION

82ND AIRBORNE troops make a night jump at Fort Bragg, in North Carolina.

RENDEZVOUS WITH DESTINY

The US Army introduced two specialist airborne divisions: the 82nd "All American" and the 101st "Screaming Eagles." They share what the first commander of the 101st, William C. Lee, called a "rendezvous with destiny." Their men and women are trained to be the first troops into any situation. They are ready to go anywhere in the world on short notice. Whether by parachute or by helicopter, the airborne get themselves to where they can hit the enemy hardest.

EYEWITNESS

"Let me call your attention to the fact that our badge is the great American eagle. This is a fitting emblem for a division that will crush its enemies by falling upon them like a thunderbolt from the skies."

Major William C. Lee
First commander, 101st Airborne

ORGANIZATION

The 82nd and the 101st are both infantry divisions of the US Army. Their soldiers carry light equipment and fight on foot, like other infantry. But the airborne arrive by air. The 82nd Division reaches its destination by jumping from aircraft with parachutes. The 101st is trained for air assault. Its troops are landed behind enemy lines by helicopter.

◄ 101ST Pathfinder keeps guard as a Black Hawk lands to retrieve his unit in Afghanistan.

ORGANIZATION

THE SHOULDER patch of the 82nd shows the letters AA, for "All American."

ALWAYS READY

The description "airborne" doesn't tell the whole story. Both divisions carry their own artillery and transportation. They have their own counterintelligence and reconnaissance units. Both divisions are part of XVIII Airborne Corps, the part of the US Army designed for rapid deployment anywhere in the world. The 82nd always has a battle-ready task force on standby that can be ready to jump and fight anywhere in the world in only 18 hours.

CHOPPERS

The helicopter transformed warfare. The first model was invented by Igor Sikorsky. It went into production in 1943, but not many were used in World War II. The first time it was used extensively was in the Vietnam War in the 1960s.

7

US AIRBORNE FORCES

82ND AIRBORNE

The 82nd started life on the ground. It was formed in 1917 as an infantry division of volunteers for the US Army in World War I. The original members of the 82nd came from all 48 states that belonged to the Union at the time. That's how it got its nickname of the "All-American" Division. The division fought on the western front in France in 1918. It was disbanded after the end of the war.

A "STICK" of paratroopers fall from an Air Force C-130 Hercules.

82ND AIRBORNE

EYEWITNESS

"What can I tell you about airborne people? I can tell you they're brothers. I never made such good friends as I made in the service. When a guy needed help he got it."

Charles Miller
Company D, 505th Brigade, 82nd Airborne

SINCE WORLD WAR II

The 82nd was reformed to fight in World War II. In August 1942 it became the first parachute division in the US Army. It saw action in the invasion of North Africa and in northern Europe. Since then, the division has served in virtually all major US conflicts around the world, including Vietnam, Iraq, and Afghanistan. It has also helped during humanitarian disasters, including Hurricane Katrina in New Orleans in 2005 and the earthquake that devastated Haiti in 2010.

AS PARATROOPERS leave the C-130 aircraft, static lines open their parachutes.

US AIRBORNE FORCES

A SOLDIER of the 101st rappels from a UH-60 Black Hawk.

101st Airborne

Since the "Screaming Eagles" were created in 1942, they have lived up to the claim by their first commander, William C. Lee, that they had a "rendezvous with destiny." The division has seen many firsts. They led the D-Day landings on June 6, 1944. They tried to liberate Holland in Operation Market Garden. And they fired the first shots of the Persian Gulf War in 1991.

101st AIRBORNE

A CH-47 Chinook drops a combat team in Afghanistan.

NEW SPECIALTY

In the Vietnam War (1963–1975) the nature of the 101st division's missions changed. In the jungles of Southeast Asia they made only one parachute jump. Instead, they were more often dropped behind enemy lines by helicopter. In 1974, the 101st ceased being a parachute regiment. It became the world's only specialist air assault division. It was trained for rapid insertion by helicopter and to fight behind enemy lines or in remote locations with few access roads.

EYEWITNESS

"I sent the 101st Airborne Division on so many important missions, never once did its fighting men fail to add new luster to their reputation as one of the finest units in the Allied Forces."

Dwight D. Eisenhower
Supreme Allied Commander, World War II, and US President

US AIRBORNE FORCES

AIR ASSAULT SCHOOL

Recruits to the 101st Airborne attend US Army Air Assault School at Fort Campbell, in Kentucky. The 10-day course is very demanding. Before training even begins, recruits have to pass a grueling obstacle course and take other tests. They also have their kit examined to make sure they've packed what they were told to bring. If anything is missing, they're out.

TRAINEES rappel down a tower. They have to stop and hang part way down.

AIR ASSAULT SCHOOL

MENTAL STRENGTH

The course covers all sides of airmobile operations. Recruits are familiarized with the aircraft they'll be flown in. They also learn to load artillery or vehicles on slings beneath aircraft. Another part of the course teaches recruits to exit a hovering helicopter by sliding down a rope. The course ends with a 12-mile (19 km) march. A candidate has to finish it within three hours to graduate and be entitled to wear the Air Assault badge.

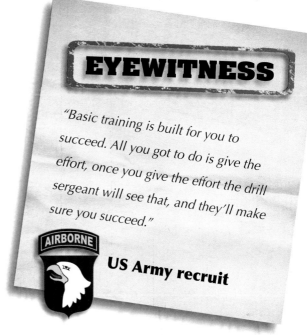

EYEWITNESS

"Basic training is built for you to succeed. All you got to do is give the effort, once you give the effort the drill sergeant will see that, and they'll make sure you succeed."

US Army recruit

AN AIRBORNE candidate takes part in an obstacle course at the Air Assault Course.

US AIRBORNE FORCES

JUMP SCHOOL

New recruits to the 82nd Airborne take a three-week Basic Airborne Training (BAT) course at Jump School, at Fort Benning, in Georgia. For the first week, called Ground Week, they don't even get into the air. They get used to the feel of wearing a parachute, and practice how to fall correctly when they land.

AIRBORNE paratroopers stage a large-scale exercise with Air Force airplanes in 2011.

JUMP SCHOOL

THE PLF

The parachute-landing fall (PLF) teaches a paratrooper to hit the ground, shift his weight, rotate his legs, and keep his feet together. If he follows the hit, shift, rotate, and feet together rule, he will walk away. If he doesn't, it could mean disaster.

INTO THE AIR

The second week is Tower Week. Trainees jump on wires from a 34-foot- (10 m) high tower to simulate the feeling of landing from a real jump. They also learn how to exit safely from an airplane. In the last week, called Jump Week, recruits make five jumps, including one at night. Not everyone makes it. Out of an average class of 360 recruits, as many as 100 don't graduate. Jump School doesn't only train the 82nd. It trains many parts of the Army, such as the Special Forces.

AN ARMY CAPTAIN trains for a door check during a jump.

15

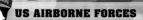

US AIRBORNE FORCES

PARACHUTE DROPS

Once an 82nd paratrooper has completed Jump School, he can jump in combat situations. Large cargo aircraft mean that hundreds or even thousands of troops can be dropped behind enemy lines on short notice. It's a long journey, which is sometimes uncomfortable, and it ends with a jump into danger.

US AND EGYPTIAN paratroopers prepare for a combined jump.

PARACHUTE DROPS

IN THE AIR

Paratroopers train to jump at intervals of just a few seconds. In the air, they form a group known as a "stick." Usually, their parachutes are opened by fixed lines as they exit the aircraft. Sometimes, though, they freefall and open their own parachutes at lower altitudes. Once the canopy is open, they have to steer to their desination and avoid hitting their colleagues. Once on the ground, the most important thing is to get out of the harness and be ready to fight.

A PARATROOPER from the 82nd Airborne comes in for landing.

EYEWITNESS

"I once heard a great Airborne leader say, 'I don't like jumping out of airplanes but I sure like being around people who do!' For me, I like both. I have always liked jumping out of perfectly good airplanes. I like the dedication, the courage, the esprit de corps of the Airborne soldier."

General Hugh Shelton
Jump School

17

US AIRBORNE FORCES

AIR ASSAULT

The Screaming Eagles are the world's only dedicated air assault division. Their first airmobile operations took place during the Vietnam War. The 101st Airborne learned how to get into and out of enemy territory quickly. The threat of attack meant that it was often too dangerous to land. The soldiers used ropes to drop to the ground. They also learned to jump into helicopters flying just above the ground.

DOOR GUNNERS on a CH-47 Chinook helicopter survey the ground as they fly in to Kandahar Airfield, in Afghanistan.

AIR ASSAULT

FAST ROPING

The fastest way to exit a helicopter without a parachute is by fast-roping. The troops slide down a rope, using friction to control their slide and to stop themselves gently. The technique comes from rock climbing.

USEFUL ALLIES

The Screaming Eagles can get behind enemy lines fast and with maximum firepower. Unlike paratroopers, who jump with only light infantry arms, air assault troops bring everything. They sling vehicles and big guns beneath huge cargo helicopters. Because the 101st are so flexible, everyone wants them. They are in great demand from commanders throughout the military who want a fast-moving, powerful strike as part of an operation.

A MEMBER OF a combat team watches for insurgents from a low-flying helicopter.

US AIRBORNE FORCES

WORLD WAR II
1939-1945

Early on June 6, 1944, called D-Day, men from the 82nd and 101st Airborne were the first Allied soldiers in occupied France. They arrived by parachute or glider to clear the way for landings on Utah and Omaha Beaches. Both divisions had been activated in 1942. They served with distinction throughout the war.

PARATROOPERS wait in a glider that will take them to France for D-Day.

WORLD WAR II

PARATROOPERS from the 101st Division land in the Netherlands during Operation Market Garden, September 17, 1944.

MARKET GARDEN

Paratroopers from the 82nd had already taken part in landings in North Africa, Sicily, and Italy. The Germans called them "devils in baggy pants." In 1944 the Airborne dropped into the Netherlands as part of Operation Market Garden. They had to hold a 16-mile-(25 km) long corridor open for an Allied advance from France. The advance failed, but the 101st held their position for 72 days despite heavy enemy fire.

EYEWITNESS

"As I stood at the door of my plane I noticed smoke coming from one of H Company's planes to my left. The plane shivered and began to fall as 17 paratroopers—an entire stick—streamed out the door. No additional chute opened. The pilot and copilot didn't make it."

T. Moffat Burriss
82nd Airborne,
Operation Market Garden

21

US AIRBORNE FORCES

BASTOGNE

1944

MEMBERS OF the 101st Airborne Division leave Bastogne at the end of the siege.

After Market Garden, the 101st was sent to Bastogne, in Belgium near the Ardennes Mountains. German forces attacked the town in December 1944 as part of the Battle of the Bulge. The airborne division was sent by road to reinforce the town. Bastogne guarded the road to the Germans' real target: the port of Antwerp.

BASTOGNE

ALLIED AIRCRAFT drop supplies to the US troops besieged in Bastogne.

UNDER SIEGE

The Americans were outnumbered and low on supplies and cold-weather gear—and it was a bitterly cold winter. Nevertheless, when the commander of the 101st received a German request to surrender, he gave a one-word reply: "Nuts!" For seven days, the Airborne fought off German Panzer units until a US armored column relieved the siege. The 101st pursued the Germans all the way to Berlin, the German capital. Meanwhile, the 82nd took Grave and Nijmegen in the Netherlands.

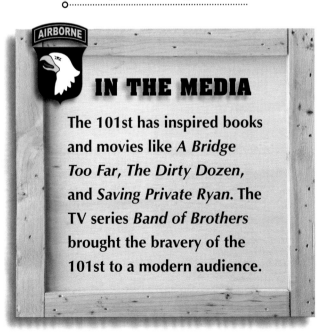

IN THE MEDIA

The 101st has inspired books and movies like *A Bridge Too Far*, *The Dirty Dozen*, and *Saving Private Ryan*. The TV series *Band of Brothers* brought the bravery of the 101st to a modern audience.

US AIRBORNE FORCES

VIETNAM
1963–1975

The 101st Airborne arrived in Vietnam in July 1965 for its most costly conflict. It would lose almost 20,000 men, twice the number killed in World War II. The 101st Airborne fought many battles in the Central Highlands and further north. In August 1968 they became a helicopter-borne air assault division.

MAP OF SOUTHEAST ASIA

UH-1 HELICOPTERS of the 101st Airborne Division fly above a smokescreen being set up for an operation.

EYEWITNESS

"The fighting had dwindled to sporadic rifle fire and an occasional grenade explosion as our infantry continued swarming over the hill. Cobra gunships roamed the skies firing mini-guns and grenades into the remaining enemy positions. The battle was ending. We had won."

Sergeant Arthur B Wiknik Jr, 101st Airborne Green Beret in Vietnam

VIETNAM

A MEDIC uses colored smoke to signal for a helicopter to evacuate casualties from Hamburger Hill.

TROOPS OF the 101st Airborne take stock in the shattered landscape after the battle.

HAMBURGER HILL

The 101st's bloodiest battle came in May 1969 at Ap Bia Mountain, better known as "Hamburger Hill." Three 101st battalions were helicoptered in to capture a North Vietnamese position on the top of the hill. The fighting was difficult. The enemies were well dug in. The Americans had to attack uphill along trails guarded by machine guns in bunkers. But with reinforcements and artillery batteries, the Americans eventually took their target after 10 days. They killed almost 700 North Vietnamese; US losses were around 72.

GRENADA
1983

The first major US military operation after Vietnam was Operation Urgent Fury. On October 23, 1983, about 7,000 US forces invaded the Caribbean island of Grenada. The US was concerned that Cuban communists were taking over the island. There were problems from the start. Forces had so little intelligence information they carried tourist maps.

MAP OF GRENADA

MEMBERS of the 82nd Airborne fire howitzers in Operation Urgent Fury.

GRENADA

FOOD TO GO

Operation Urgent Fury saw paratroopers issued with MREs (meals ready to eat) for the first time. These rations were used when it was not possible to prepare fresh food. The packaging had to withstand being dropped by parachute.

SUCCESSFUL OPERATION

The 82nd Airborne deployed to Grenada within 17 hours of receiving notice of the invasion. The infantry's role was to back up other US forces, including the special forces. The paratroopers successfully took Port Salinas airport and moved north. As they traveled through the island, local people welcomed them. Despite the lack of preparation, the mission succeeded. It was over within two months.

AN AIRBORNE patrol investigates a house in the Grenadan countryside.

US AIRBORNE FORCES

Persian Gulf
1991

82ND AIRBORNE soldiers examine a Hind Mi-24 helicopter abandoned by retreating Iraqis.

MAP OF SAUDI ARABIA

US AND COALITION forces assembled in Saudi Arabia, near the border of Kuwait.

After the Iraqi leader Saddam Hussein invaded Kuwait in August 1990, the 82nd and 101st Airborne were sent to Saudi Arabia as part of Operation Desert Shield. They helped make sure Hussein did not enter Saudi Arabia. The speed of the 82nd Airborne's deployment was vital to getting troops on the ground.

PERSIAN GULF

INTO KUWAIT

When the US-led Coalition went into Kuwait in January 1991, the 101st Airborne fired the first shots of the war. Eight 101st Apache AH-64 helicopters destroyed two Iraqi early warning radar sites. During the 100-hour war, the 18,000 soldiers and 400 helicopters of the 101st Airborne carried out airmobile attacks along the Euphrates River valley. The mission was successful and by May 1991 the 101st Airborne were home.

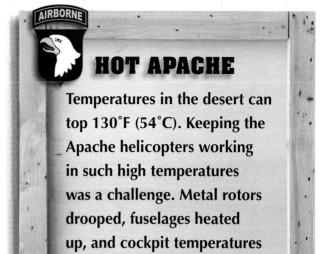

HOT APACHE

Temperatures in the desert can top 130°F (54°C). Keeping the Apache helicopters working in such high temperatures was a challenge. Metal rotors drooped, fuselages heated up, and cockpit temperatures became unbearable.

1st BRIGADE of the 101st Division await orders to move into Iraq.

US AIRBORNE FORCES

IRAQ
2003

When the US went to war with Iraq in 2003, both the 82nd and 101st Airborne were deployed. The 101st supported the 3rd Infantry Division in the invasion. It took the lead in capturing Saddam International Airport, in Baghdad. The 101st was based in Mosul for its first tour of duty.

MAP OF IRAQ

TROOPS FROM the Airborne Field Artillery Regiment fire a lightweight howitzer in Iraq.

IRAQ

SECOND DEPLOYMENT

The 101st was sent back to Iraq in 2005 to fight growing lawlessness. It had responsibility over much of northern Iraq. The division captured more than 500 insurgents. In 2006 it launched Operation Swarmer, a large-scale air assault to capture insurgents in the city of Samarra. It also trained Iraqi security forces to be able to take over when US forces left, and also helped carry out 5,000 projects to reconstruct Iraqi infrastructure.

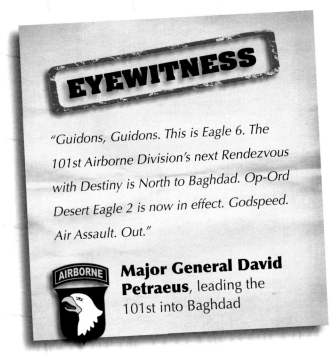

EYEWITNESS

"Guidons, Guidons. This is Eagle 6. The 101st Airborne Division's next Rendezvous with Destiny is North to Baghdad. Op-Ord Desert Eagle 2 is now in effect. Godspeed. Air Assault. Out."

Major General David Petraeus, leading the 101st into Baghdad

AN 82ND soldier uses his combat optical sight to look for threats during a night patrol in Baghdad.

AFGHANISTAN
2003

Following the terrorist attacks of 9/11, Airborne troops were sent to Afghanistan as part of Operation Enduring Freedom. Their job was to find Osama bin Laden and other al-Qaeda terrorists thought to be hiding there. Third Brigade, 101st Airborne, took part with special forces in Operation Anaconda in March 2002. In the Shah-e-Kot Valley they fought the largest battle to date with Taliban fighters.

TROOPS FROM the 101st Airborne wait to board a CH-47 Chinook for an operation in Afghanistan in 2002.

AFGHANISTAN

MAP OF AFGHANISTAN

BACK TO AFGHANISTAN

Both the airborne divisions have been sent back to Afghanistan for further tours of duty. They maintain firebases in the Ghanzi and Kandahar provinces and patrol the local areas. But they also work increasingly with the Afghans. Soldiers from 4th Brigade, 82nd Division, for example, went to Afghanistan in 2012. They learned Pashtun, the local language, so they could talk to locals and even to their enemy, the Taliban. By then, many people had begun to think that the war would eventually be ended by negotiation with the Taliban.

EYEWITNESS

"Getting more soldiers on the ground is important. The more area that you can cover, the longer you can be out there. The focus is protecting the (civilian) population, and if you have more men on the ground you can do that."

Captain Henry Hansen,
101st Airborne

AIRBORNE TROOPS prepare to search a shelter for signs of Taliban activity.

33

US AIRBORNE FORCES

HURRICANE KATRINA

2005

After Hurricane Katrina made landfall in Louisiana on August 29, 2005, the 82nd Airborne was called in as part of Task Force Katrina. New Orleans was flooded. Many thousands of people were stranded. Supplies were running out. There was a danger of disease spreading. Law and order was breaking down. Looters were stealing goods from stores.

A SPECIALIST from the 82nd Airborne gives a civilian a tetanus shot in a school classroom.

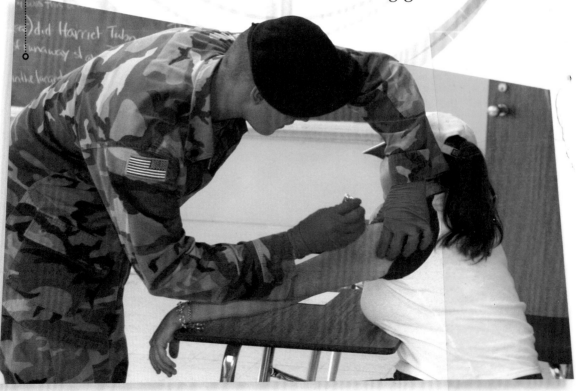

34

DISASTER FORCE

Since the Great Chicago Fire of 1871, the army has responded to most major US disasters. Only the military is able to get on the ground quickly, restore communications, organize rescue and recovery, and keep law and order.

HOMELAND OPERATION

The commander of the 82nd, Major William Caldwell, was told, "Your job is to fix the airport and fix New Orleans." About 5,000 men arrived just seven hours after the deployment order. Their jobs varied. Some went house-to-house, carrying out search and rescue. They used inflatable boats to travel through the flooded streets and pick up people stuck in their homes. Helicopters dropped food and water supplies. Armed soldiers also patrolled the streets to stop looting and to restore law and order.

82ND AIRBORNE and the Coast Guard use a Zodiac inflatable to search for survivors.

US AIRBORNE FORCES

HAITI
2010

On January 12, 2010, the Caribbean country of Haiti was devastated by a huge earthquake. Within 72 hours, the 82nd Airborne was landing at the main airport in Port-au-Prince to take charge of a relief operation. Many thousands of Haitians were left homeless and short of food. There was a possibility that disease would spread without good sanitation.

MAP OF HAITI

A SOLDIER from the 82nd Airborne leads a boy through a camp for survivors of the earthquake, looking for the child's family.

MEN OF THE 82nd unload emergency supplies after the 2010 earthquake in Haiti.

NEW MISSION

In 1994 the 82nd was deployed to invade Haiti after a coup. But while they were in the air, the military regime backed down. The 82nd landed to find that they were no longer invaders. They had become peacekeepers.

TRANSFER OF POWER

Within days planes brought around 3,500 airborne troops to Haiti. They traveled crammed among military vehicles and supplies such as water. In Haiti, many of the troops had to sleep in the open, with no tents, water, or electricity. The 82nd Airborne were often the first people in uniform to reach an area after the quake. They handed out hundreds of thousands of bottles of water and packaged meals. They also provided security to make sure emergency supplies reached the people who needed them rather than being seized by looters.

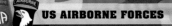

US AIRBORNE FORCES

Aircraft

A US AIR FORCE C-17 Globemaster III takes off.

The All Americans, the 82nd Airborne, make their parachute drops from large fixed-wing aircraft like the C-130 Hercules. These large transport aircraft can carry as many as 200 paratroopers at a time, but not with much in the way of comfort or luxury. The C-17 Globemaster III can transport helicopters as well as personnel. It is used to get troops and any equipment they might need into the combat zone.

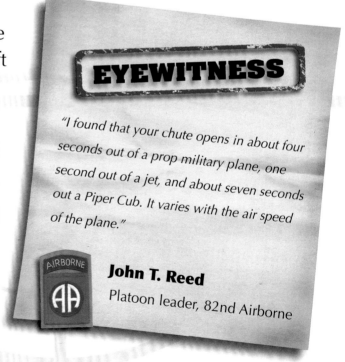

EYEWITNESS

"I found that your chute opens in about four seconds out of a prop military plane, one second out of a jet, and about seven seconds out a Piper Cub. It varies with the air speed of the plane."

John T. Reed
Platoon leader, 82nd Airborne

AIRCRAFT

VARIED HISTORY

The 82nd have used many different aircraft during their history. In World War II, they jumped from propeller planes such as the Fairchild C-119 Flying Boxcar. In Vietnam, the Bell UH-1 Iroquois helicopter, nicknamed the Huey, was used to fly troops across the dense jungle, but parachute jumps were rare. The 82nd still use helicopters for parachute drops but fixed wing aircraft can transport far more troops. This is important when the division is deployed as a rapid-response unit and needs to hit the ground quickly and in force.

SOLDIERS OF THE 82ND Airborne carry their gear during a flight on a C-17 Globemaster.

US AIRBORNE FORCES

HELICOPTERS

Helicopters are the lifeblood of the Screaming Eagles, the only helicopter division in the US Army. Since 1968 they have used helicopters for air assault operations. Helicopters were more flexible than fixed wing aircraft in the thick jungles of Vietnam. They are vital in the terrain of Afghanistan for moving troops where there are few roads.

MEMBERS OF the 101st Airborne use slings to attach a howitzer beneath a UH-60 Black Hawk.

HELICOPTERS

A 101st AIR assault from a UH-1 Huey Iroquois helicopter.

VARIED ROLES

The 101st Airborne has more helicopters than any other division in the US Army: 281 in early 2012. They include UH-60A Black Hawks and AH-64 Apaches. Black Hawks take troops to and from the combat zone. The Apache gunship is armed with Hellfire missiles and is used for ground attack and for surveillance. Other helicopters used by the 101st Airborne include the OH-58C Kiowa, which was deployed during the Persian Gulf War to provide air-to-air protection.

EYEWITNESS

"The mission of the 64 (Apache) is to find, fix, and destroy the enemy. [The Apache] is designed to destroy tanks, day or night."

Lt. Col. Scott Hasken
Commander, 1st Battalion, 101st Airborne

Special Gear

The 82nd and 101st Airborne are light infantry air divisions. As a rule, that means other units have to bring along the big tanks and artillery. When the 82nd deploys, its job is to get to the action first. Parachuting in means the 82nd can't carry much equipment. Later units bring in vehicles and big guns.

Little Things

Some of the most valuable gear an airborne soldier can carry is stuff you can buy in the drugstore, such as sunscreen and chapstick. Good sunglasses and kneepads can also make a big difference in the field.

SOLDIERS of the 82nd Airborne fire a howitzer during an exercise.

SPECIAL GEAR

AN 82ND AIRBORNE armored crew drive an M551 Sheridan light tank, painted in camouflage, for an exercise.

AIRBORNE VEHICLES

The airborne still need vehicles to do their job. Helicopters fly in four-wheel drive High Mobility Multipurpose Wheeled Vehicles, or Humvees. The Humvees are carried by sling or inside transport aircraft. They can go offroad, which helps avoid the threat of roadside bombs. Their tires have to be good, so they can avoid punctures or becoming stuck in difficult terrain.

US AIRBORNE FORCES

WEAPONS

Airborne troops don't carry too many weapons. That's especially true for paratroopers jumping from aircraft. They often restrict themselves to a knife and a gun. They used to carry purpose-built airborne guns that could be taken apart. Today, a paratrooper is more likely to carry an M-16 or another kind of multipurpose machine gun.

MEN FROM the 101st carry their kit from a Black Hawk helicopter on a mission in Iraq.

WEAPONS

MACHINE GUNS

For the air-assault 101st Division, the machine gun is standard issue. In combat situations like Afghanistan, where much of the fighting is done at a distance, a sniper rifle is more useful, because it has a much longer range. Other weapons used by the Airborne include shotguns, rifles, grenades, artillery guns, and 20mm antiaircraft cannons. They have to be maneuvered into place on small utility vehicles.

SPECIAL GUNS

Paratroopers have to be careful as they land. Their rifles might fire accidentally. In World War II, some paratroopers jumped with their rifles in pieces. They had to reassemble them on the ground, losing precious seconds before they could start fighting.

AIRBORNE TROOPS in Iraq carrying M4 assault rifles prepare to search a property.

GLOSSARY

air assault (EHR eh-SOLT) A military attack carried out by soldiers in helicopters.

airmobile (ehr-MOH-bul) Operations in which troops are moved by air.

artillery (ahr-TIH-lur-ee) Guns that fire large-caliber shells, like cannons and howitzers.

canopy (KA-nuh-pee) The part of a parachute that opens.

deployment (dih-PLOY-ment) Putting troops into position for a particular purpose.

division (dih-VIH-zhun) A large, self-contained part of an army that can operate on its own.

glider (GLY-der) An aircraft without an engine that flies by using air currents.

howitzer (HOW-it-ser) A type of cannon with a short barrel.

humanitarian (hyoo-ma-nuh-TER-ee-un) To do with the welfare of many people.

insertion (n-SER-shun) Getting troops into hostile territory.

insurgent (in-SER-jint) A fighter against the established government who uses guerrilla or terrorist tactics.

intelligence (in-TEH-luh-jents) Any information found out about the enemy.

parachute (PAR-uh-shoot) A cloth pouch that slows someone or something as it falls through the air.

reconnaissance (rih-KAH-nih-zents) A military survey of ground held by the enemy.

recruit (rih-KROOT) A new member of a group.

rendezvous (RON-dih-voo) A meeting at a set time and place.

siege (SEEJ) Surrounding a place, such as a town or fort, to force it to surrender.

sling (SLING) Looped straps of rope or chain used to carry loads beneath aircraft.

stick (STIK) A group of paratroopers who jump from an airplane in a column.

FURTHER READING

Adams, Simon. *World War II*. New York: DK Children, 2007.

Benoit, Peter. *Hurricane Katrina*. A True Book. Danbury, CT: Children's Press, 2012.

Gifford, Clive. *Why Did the Vietnam War Happen?*. Monments in History. New York: Gareth Stevens, 2011.

Jackson, Kay. *Military Helicopters in Action*. Amazing Military Vehicles. New York: PowerKids Press, 2009.

WEBSITES

Due to the changing nature of Internet links, PowerKids Press has developed an online list of websites related to the subject of this book. This site is updated regularly. Please use this link to access the list:
www.powerkidslinks.com/usf/air/

INDEX

101st Airborne 5, 10–13, 18–19, 22–25, 30–31, 40–41
82nd Airborne 5, 7–9, 13–14, 16–17, 27, 30, 34–38

Afghanistan 18, 32–33
AH-64 Apache 29, 41
air assault 11, 18–19, 45
Air Assault School 12–13
aircraft 38–39
All Americans 5, 7–8

Bastogne, Belgium 22–23

C-130 Hercules 8–9, 38
C-17 Globemaster 38–39
Caldwell, William 35
CH-47 Chinook 11, 18, 32

D-Day 4, 20

equipment 38–45

Grenada 26–27
guns 44–45

Haiti 36–37
Hamburger Hill 25
helicopters 7, 11, 19, 24, 40–41
history 20–37
howitzers 30, 42
Humvees 43
Hurricane Katrina 34–35
Hussein, Saddam 28

Iraq 28–31

Jump School 13–14

Kuwait 28–29

Netherlands 21, 23
New Orleans, Louisiana 34–35

Operation Anaconda 32
Operation Enduring Freedom 32–33
Operation Market Garden 21
Operation Swarmer 31
Operation Urgent Fury 26–27
organization 6–7

parachute drops 16–17
parachute-landing fall (PLF) 15
paratroopers 4, 14, 16–17
Persian Gulf War 28–29
Petraeus, David 31

rifles 44–45

Screaming Eagles 5, 10, 18–19, 40–41

Taliban 32–33
training 12–15

UH-1 Huey 24, 39, 41
UH-60 Black Hawk 6, 10, 40–41, 44

Vietnam War 11, 18, 24–25

weapons 44–45
World War I 4, 8
World War II 4, 9, 20–23, 39, 45